D1060385

BUFFALO

Jen Green

Grolier
an imprint of

www.scholastic.com/librarypublishing

Published 2009 by Grolier
An Imprint of Scholastic Library Publishing
Old Sherman Turnpike
Danbury, Connecticut 06816

For The Brown Reference Group plc
Project Editor: Jolyon Goddard
Picture Researchers: Clare Newman, Sophie Mortimer
Designer: Sarah Williams
Managing Editor: Tim Harris

Volume ISBN-13: 978-0-7172-8037-7
Volume ISBN-10: 0-7172-8037-3

Library of Congress Cataloging-in-Publication Data

Nature's children. Set 5.
p. cm.
Includes index.
ISBN-13: 978-0-7172-8084-1
ISBN-10: 0-7172-8084-5 (set)
1. Animals--Encyclopedias, Juvenile. I. Grolier Educational (Firm)
QL49.N386 2009
590.3--dc22
2008014674

Printed and bound in China

PICTURE CREDITS

Front Cover: **Shutterstock**: Johan Swanepoel.

Back Cover: **Shutterstock**: EcoPrint, TAOLMOR, Jason Vandehey, Stephanie Van der Vinden.

Alamy: Danita Delimont 21, 46, Khmer Images 9, Roy Lawe 17, Chris Wilson 14; **Corbis**: Marilyn Angel Wynn 41; **FLPA**: Stefan Auth 30, Nigel Cattlin 13; **NaturePL**: Pete Oxford 5; **Photolibrary**: Michael S. Bisceglie 45, Lauree Feldman 29, Robert Harding Imagery 42; **Shutterstock**: Sebastian Burel 38, Eric Isselee 4, Jean-Michel Olives 6, Bill Kennedy 10, Dimitry Pichugin 26–27, A.L. Spangler 37, Johan Swanepoel 34, Gleb Vinnikov 33, Neil Wigmore 2–3, Bruce Yeung 22; **Still Pictures**: Zoltan Nagy 18.

Contents

FACT FILE: Buffalo

Class	Mammals (Mammalia)
Order	Cloven-hoofed mammals (Artiodactyla)
Family	Antelope, cattle, sheep, and goats (Bovidae)
Genera	*Bubalus*, *Bos*, *Bison*, and *Syncerus*
Species	Water buffalo (*Bubalus arnee*); yak (*Bos mutans*); African buffalo (*Syncerus caffer*); American bison (*Bison bison*)
World distribution	Water buffalo and yak live in Asia; African buffalo are found in Africa, south of the Sahara; American bison live in North America
Habitat	Water buffalo live in hot, wet climates; yak live in mountains; African buffalo and American bison live in grasslands and woodlands
Distinctive physical characteristics	Large mammals with a big head, shoulder hump, cloven hooves, and horns; color of coat varies from white and gray to brown and black
Habits	Grazing animals that chew the cud; live in herds in the wild
Diet	Grass and other plants

Introduction

The word *buffalo* is used to describe a variety of different—but related—types of **bovids**. Bovids are **mammals** and members of the cattle family.

Water buffalo have large, curving horns. Most water buffalo are **domestic** animals that work in the fields of many Asian countries. Their close relative, the yak, is found in mountain areas of Asia. Like water buffalo, most yak are raised by humans. The African buffalo is a more distant relative and has never been tamed by humans. The shaggy-haired bison of North America are often incorrectly referred to as buffalo, too.

Wild water buffalo are very rare.

The owner of this domestic water buffalo keeps it on a leash so that it cannot wander off.

The Buffalo Family

The large and varied family of bovids has about 140 **species**. Asian and African buffalo, yak, and domestic cattle are all members of the family. It also includes the bison, which is the correct name for the North American buffalo. Sheep, goats, and antelopes are distant relatives, too.

All bovids have certain things in common. They all have **cloven**, or split, **hooves**. Both male and female bovids have horns set high on their head. Horns are useful weapons if danger threatens. Unlike the **antlers** of deer, a buffalo's horns do not branch. Nor are they shed each year, like deer antlers. Buffalo and their relatives live in groups called **herds**. The herd is usually made up of adult females and their young. The males live separately and join the herd when its time to breed.

Chewing the Cud

Buffalo and their relatives are all plant eaters. Plants contain a tough substance called **cellulose**, which is hard to digest, or break down. Like all bovids, buffalo have a complicated **digestive system** with four stomach chambers to break down their stringy food.

Buffalo and bison mostly **graze** on grass and meadow plants. They tear off grass with their long, flexible tongue and front teeth. After chewing it a little, they swallow. The food passes into the first stomach chamber. Later, usually when the animals are resting, the food returns to the mouth for more chewing. This is called chewing the **cud** or **ruminating**. The animals that do that are called ruminants. After a good chewing, the food is swallowed again. This time it passes through all four chambers of the stomach and through the rest of the digestive system. Chewing the cud allows the animals to extract as many nutrients as possible from their tough plant food.

A water buffalo, like many other grazers, has microscopic germs called bacteria in its stomach. These germs help break down tough plant matter.

Water buffalo have long, sparse hair that ranges from ashy gray to black.

Water Buffalo

The water buffalo, also known as the Asian buffalo, is an extremely hefty animal. A large male, or **bull**, can stand 6 feet (1.8 m) tall at the shoulders, measure 9 feet (2.7 m) from head to backside, have a tail 3⅓ feet (1 m) long, and weigh up to 1 ton (1 tonne)! This buffalo also has very wide horns. From tip to tip, the huge, curving horns can measure 6½ feet (2 m) across! Wild water buffalo are fearsome creatures. Even a tiger thinks twice before attacking. Water buffalo **calves** are more at risk from **predators**, but the fierce mothers defend their babies with their horns. When not in danger, water buffalo are naturally quiet and gentle animals. They live near water and **wallow** in muddy pools to keep cool at midday. Water buffalo mainly feed at night and spend their days dozing and chewing the cud.

Swamp and River

There are not many water buffalo left in the wild. Today, these large beasts are mainly domestic animals, reared for meat, milk, and to help with farmwork. There are two main types of domestic water buffalo: river and swamp buffalo.

The river buffalo is the larger type, weighing 2,000 pounds (900 kg) or more. It is dark gray in color. The large horns curve in a circle and have deep "wrinkles," or grooves. This buffalo is raised mainly for its meat and milk. It lives in India and Southeast Asia.

The swamp buffalo is found in China, the Philippines, and Southeast Asia. It is a paler shade of gray, and weighs up to 1,600 pounds (720 kg). This type of water buffalo has large horns that often stick out sideways. It also has V-shaped markings on its chest, similar to a sergeant's stripes. Swamp buffalo are commonly used as work animals. They also provide meat and milk.

A large herd of river buffalo wallows in a water hole in India.

A Japanese farmer uses a water buffalo to crush sugarcane.

Working Animals

Humans have been raising water buffalo and working with them for thousands of years. Strong and sturdy, water buffalo are used to pull a plow, haul logs, or even heave tree stumps out of the ground.

These beasts are also used to power water pumps and machines, such as simple mills, that grind grain into flour. The buffalo usually walk around in a circle to work the machinery. They are also used to pull wagons. With their wide hooves, they are able to keep their footing both in muddy rice fields and on narrow, stony trails.

It's no wonder that many Asian farmers prefer their water buffalo to a tractor. After all, tractors need fuel and can break down. Buffalo need no fuel other than grass, and they rarely get sick or tired!

Dependable Beasts

Water buffalo are extremely useful as domestic animals. In addition to being very powerful, they are also gentle—so gentle that even a young child can handle them. They are usually very healthy and hardly ever catch diseases. As plant eaters, they cost little to feed, and they live for up to 35 years.

Water buffalo are easy to breed. The females, or **cows**, are ready to have young from the age of two years. A cow will produce a calf every other year for at least 16 years. Female water buffalo rarely have problems giving birth. They produce more than enough milk for their offspring—enough for people, too.

The water buffalo is so dependable—and so important to Chinese culture—that the Chinese even have a saying about the animal. "If I die, you will weep, but if the water buffalo dies, you will starve."

Water buffalo cows are pregnant for about 11 months. The calf is born with a yellowy or reddish-brown coat.

Mozzarella, an Italian cheese, is traditionally made from water buffalo's milk.

Milk and Cheese

In addition to being a useful work animal, water buffalo also produce milk for human consumption. Throughout Asia, most people drink buffalo's milk, not cow's milk. They also use it to make butter and other dairy produce, such as yogurt and ice cream.

In Europe, especially in Italy, buffalo's milk has another use. It is used to make cheese. In fact, mozzarella cheese, which is used to top pizzas, is traditionally made from buffalo's milk. Today, most of the mozzarella sold in supermarkets is made from cow's milk. But for an expert Italian cook, only the real thing will do—creamy mozzarella made from white buffalo's milk!

Nothing Wasted

The trusty water buffalo continues to be useful even when its life is over. The meat from these beasts is highly prized in many parts of the world. Water buffalo meat is similar in taste to beef. However, it is healthier to eat than beef. That's because it contains less harmful **cholesterol** (KUH-LESS-TUH-ROLL). If eaten in high amounts, cholesterol can lead to heart disease.

In some places, water buffalo are raised mainly for their meat. But other parts of the animal are also useful. Their skin is used to make leather. Long-wearing, soft buffalo leather is in great demand in some places. The horns are also used to make products such as knife handles, so little of the animal is wasted!

Water buffalo horns are used to decorate a temple in Indonesia.

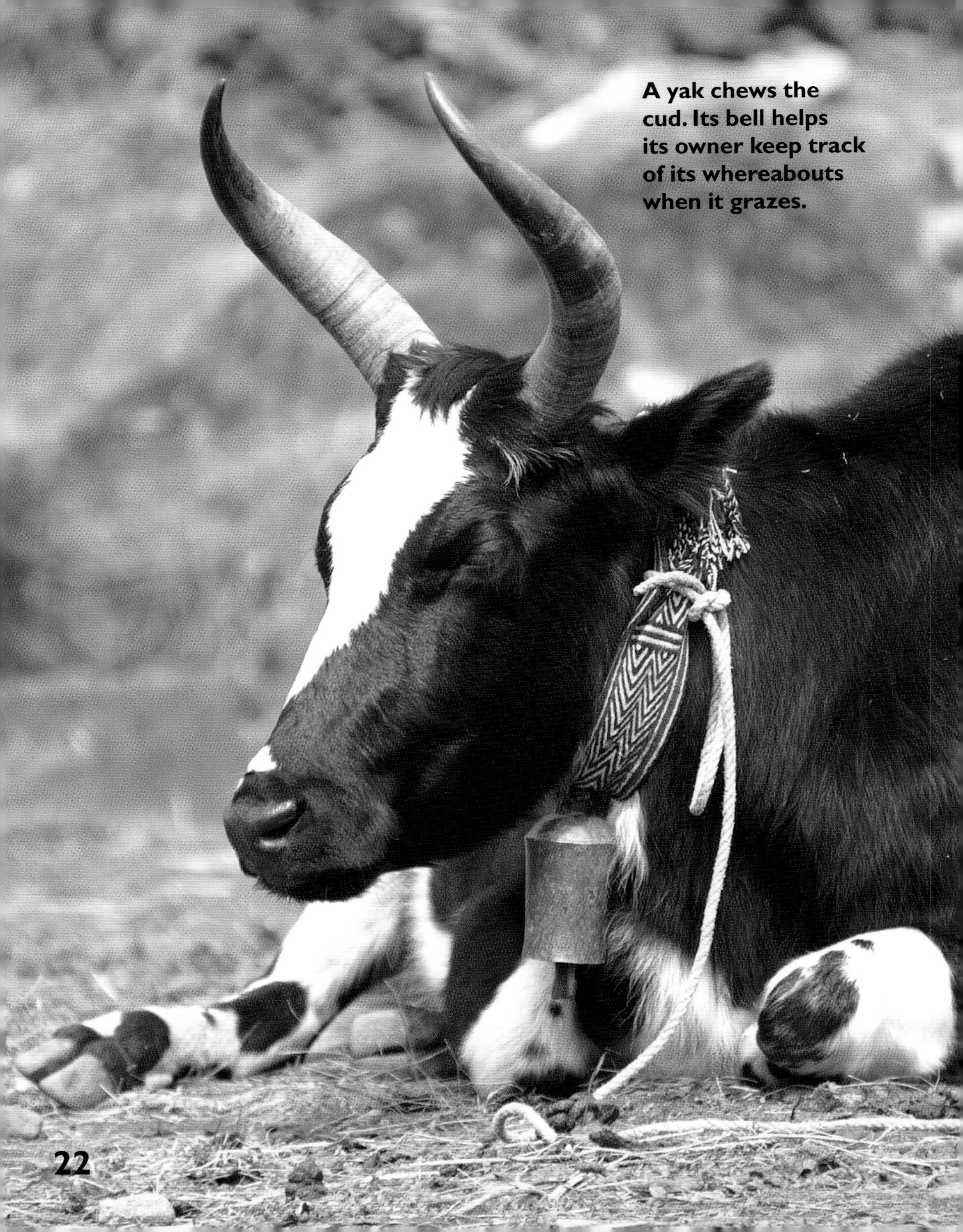

A yak chews the cud. Its bell helps its owner keep track of its whereabouts when it grazes.

The Mighty Yak

The yak is a close relative of the water buffalo. However, it comes from very different surroundings. The yak's natural **habitat** is high in the mountains of central Asia, including the Himalayas. Like water buffalo, these large beasts have been domesticated for at least 2,000 years. There are about 12 million domestic yak in Asia alone. But there are very few wild yak left today.

One of the first things noticed about yak is their size. These rugged beasts stand up to 6 feet (1.8 m) tall at the shoulders. Their long, dark, shaggy hair makes them look even bigger.

Like other bovids, yak are grazers. They naturally like to graze in the early morning and evening. Wild yak spend the remainder of the day resting and chewing the cud. However, domestic yak have a lot of work to do!

Mountain Life

The mountain climate where yak live is harsh. It is bitterly cold in winter and cool even in summer. However, yak are well suited to their surroundings. Their enormous bulk helps keep them warm. These beasts have two layers of hair. The outer coat of long, coarse hair keeps out snow and the whistling, chilly winds. An undercoat of fine, dense hair keeps the yak cozy in subzero temperatures. As summer approaches, the thick coat is mostly shed in clumps. That makes yak look pretty scruffy in spring!

Yak are sturdy beasts with short legs and large, rounded hooves. These skilled climbers are also surprisingly graceful. A yak thinks nothing of sliding down a slippery, ice-covered slope that most other animals would not go near. It will also pick its way up a narrow rocky path or swim across an icy river.

Life in Thin Air

Wild yak live at heights of more than 16,000 feet (4,870 m) in the Himalayas. At that height, the air is "thin." That means it contains much less oxygen than the air at sea level. Mountaineers who climb in high places often have trouble getting enough oxygen into their body. Some have difficulty breathing. For those reasons, many mountaineers use bottled oxygen so they can keep climbing.

Yak don't need bottled oxygen! To survive at these heights they have a special kind of blood that can absorb more oxygen than the blood of other animals. Yak are so well adjusted to the high life that they are often used to carry equipment to high camps on mountains for use by human mountaineers!

A herd of domestic yak grazes in Mongolia.

Hunting Yak

For hundreds of years, the people of central Asia have hunted wild yak. Yak meat was always highly prized. Like water buffalo meat, yak meat is low in fat and cholesterol, which makes it a fairly healthy meat to eat.

The skin of the wild yak was also prized. Hard-wearing yak leather was used to make saddles, tents, whips, shoes, and other objects. Bulky and not very swift, yak made an easy target for the hunters. When rifles reached central Asia, killing wild yak became even easier. Even the yak's fierce charge could not help save it. Today, after hundreds of years of hunting, few wild yak remain roaming the mountains of Central Asia.

Local people have placed this painted yak skull in a mountain pass in the Himalayas.

As pack animals, yak can cover more than 20 miles (32 km) each day.

The Useful Yak

The domestic yak is a very useful beast. In Tibet, which is now part of China, people sometimes ride on yak. They are also often used to deliver the mail!

Yak make excellent pack animals. Big and strong, these animals are able to carry very heavy loads. Their broad hooves help make them sure-footed. Being thoroughly at home in high altitudes, yak are able to cross mountain passes of more than 17,000 feet (5,200 m) high. No other animal or motor vehicle can make it across some of the high, snowy passes that yak cross with ease!

Treasure of Tibet

The people of Tibet are very proud of their domestic yak. In addition to providing transportation, the female yak, called a **dri**, also produces rich milk. The milk is used to make butter, cheese, and other foods—all of which Tibetans depend upon to survive. Yak-butter tea is widely consumed. It has a very unusual taste—it is salty, not sweet. A dri produces so much milk that she can be milked three times a day.

In addition to providing milk, the yak's soft, woolly underfur is woven to make clothes and blankets. The long, coarse outercoat is also useful. It can be woven into hard-wearing tents and mats. The yak's long tail was once used as a tassel to decorate the helmets of senior Tibetan officials! It's no wonder that these animals have a special place in the history and culture of Tibet.

Even the dung of yak is used by Tibetans. It is burned as fuel.

Unlike the water buffalo, the African buffalo has never been domesticated.

The African Buffalo

African buffalo live in the nondesert parts of Africa. Although African buffalo and water buffalo share the name "buffalo," they are not close relatives. African buffalo are smaller than water buffalo or yak. African buffalo bulls are larger than the cows. They also grow thicker horns than the cows.

African buffalo live in various habitats, including swamps, grasslands, and forests, from sea level to high up in the mountains. The cows and calves live in large herds. Young bulls often form groups of three or four. As they become older, they tend to live on their own. However, during the **breeding season**, bulls briefly join the herd in order to **mate** with the cows.

Buffalo's main enemies, other than human hunters, are lions, leopards, and hyenas. An adult buffalo can defend itself against, and even kill, a lion. It usually takes more than one lion to bring down an adult buffalo. Hyenas and leopards mainly hunt and eat calves. The herd usually defends any member that is attacked.

American Bison

The American buffalo is not really a buffalo at all. This famous creature, a symbol of the West, is actually a bison. According to scientists, the Asian water buffalo and the African buffalo are the only true buffalo. French explorers in the Americas were probably the first Europeans to see these large, shaggy beasts. They called them *boeufs* (BOOFS), which means "oxen" in French. As time passed, the word *boeuf* became "buff" in English, and eventually "buffalo." The name is now so much a part of American culture, that the bison will probably always be called a buffalo.

This unmistakable creature has a large head, shoulder hump, black horns, and a long, straggly "beard." Big males stand up to 6 feet (1.8 m) tall at the shoulders, and can weigh 3,000 pounds (1,350 kg). Long, shaggy hair covers the animal's head, shoulders, and front legs.

American bison originally came from Asia. They crossed from Siberia to Alaska thousands of years ago, using a natural land bridge that existed at that time.

A herd of bison prepares to cross a river in Yellowstone National Park, in Wyoming.

Bison Herds

Like other bovids, American bison live in herds. Groups of cows and their calves are led by an experienced female. In the breeding season, a few bulls join the herd. They stage fierce head-butting contests to win the right to mate with the cows. Females give birth to a single calf nine months after mating. The baby is quickly on its feet, and then the mother and young rejoin the herd.

The bison is a tough creature that can cope with freezing winters. It pushes deep snow away with its head to reach the grass underneath. A thick winter coat keeps it warm. In spring, the heavy coat is shed. In summer, these beasts wallow in marshes or dusty hollows to escape flies. As fall approaches, the shaggy coat grows again to prepare the animal for the cold winter.

A Way of Life

For thousands of years, Plains Indians such as the Pawnee, Sioux, and Cheyenne depended on the bison. This animal gave them almost everything they needed to survive. Bison meat was a staple food, eaten either fresh or dried. Bison skin was used to make clothing, tipi covers, ropes, boats, shields, and even coffins. Bison hair was braided into rope, while the horns were used as drinking cups.

Even the bison's bones were put to use. They could be carved to make bows, arrowheads, and even children's toys. The ribs were used to form the runners of dogsleds. Even the dried bison dung was used as fuel for the campfire. It is easy to see why the Plains Indians honored the bison in dance and song before a hunt.

A Sioux warrior's shield depicts a bison.

This animal is a European bison. It is closely related to the American bison and is now extremely rare.

Mass Slaughter

Before Europeans arrived in North America, Plains Indians hunted bison on foot. In the 1500s, Europeans brought horses, which Native American tribes were soon using to chase the bison. The Europeans also brought with them the rifle, which made hunting simpler. However, Plains Indians killed only the animals they needed for survival. They never took many bison, even though the herds were huge. Experts estimate that in 1851 more than 50 million bison roamed the Great Plains.

The arrival of European settlers resulted in disaster for the bison. The newcomers were soon slaughtering, not thousands, but millions of bison every year. Some were shot for food, others for their skin. Many were killed just for sport. By 1865, there were just 15 million bison left. But the worst was still to come. The settlers were in competition with the Native Americans for land. They killed the bison to clear land for farming and to remove their rivals' source of food. By 1890, fewer than 1,000 plains bison were left.

Saving the Bison

Luckily the story of the bison has a happy ending. In 1894, just in the nick of time, the United States Congress passed a law that made killing bison illegal. Small herds still survived on a few ranches, and also in what was to become Yellowstone National Park.

In 1905, President Theodore Roosevelt, who had lived in the West, was determined to save one of North America's most famous animals. He helped to found the American Bison Society, which worked tirelessly to save the species from **extinction**. The bison has made a remarkable comeback. Today there are 120,000 bison, either living on ranches or roaming free in state and national parks.

A bison calf's coat darkens to brown as it becomes older.

Cowhands round up a herd of bison on a ranch in Montana.

Bison Ranches

In the 1860s, at the height of the bison slaughter, the animals were mostly killed for their skin or for sport. Their dead bodies were left to rot on the ground. Today, however, it is the bison's meat that has led it to be reared on ranches as a domestic animal. There are now more than 1,000 bison ranches in the United States alone.

Bison meat is high in protein, but fairly low in fat, cholesterol, and calories. That makes it healthier than most other red meats. Bison ranching is becoming increasingly popular. Bison are fairly easy to raise, because they are so well suited to life on the North American plains. Additionally, there is a lot of money to be made from selling bison meat.

Wood Bison

There are two kinds of American bison. They are the plains bison of North America's wide open spaces and the wood bison, an even larger animal that lives in the forests of Alberta, British Columbia, and the Northwest Territories in Canada. Wood bison were never as common as their plains relatives. Probably no more than 170,000 wood bison roamed the northern forests in about 1800.

During the 1800s, wood bison were slaughtered like their plains relatives. By 1957, just 200 animals remained. However, the Canadian government was determined to save the wood bison. In 1922, Wood Buffalo National Park was created—one of the largest parks in the world. By the 1990s, the numbers of wood bison had risen to 3,000. The animal is still at risk from diseases spread from domestic cattle, such as **tuberculosis**. But with luck, they will continue to roam in the wild. Like plains bison, wood bison are now also raised on ranches.

Words to Know

Antlers	The branching horns of deer, which are shed every year.
Bovids	The name used to describe the cattle family, which includes bison, buffalo, and yak.
Breeding season	The time of the year when animals come together to produce young.
Bull	An adult male buffalo or bison.
Calves	Young buffalo or bison.
Cellulose	A tough substance found in plants, which makes them difficult to digest.
Cholesterol	A fatty substance found in animal tissues.
Cloven	Describing hooves that are divided, or split, into two.
Cows	Adult female buffalo and bison.
Cud	A pasty mass of partly digested food.
Digestive system	The parts of the body involved in breaking down food. It includes the mouth, stomach, and intestines.

Domestic	Tame or raised by humans.
Dri	A female yak.
Extinction	When all of a type of animal dies and is gone forever.
Graze	To eat grass.
Habitat	The type of place where an animal or plant lives.
Herds	Groups of buffalo, yak, or bison.
Hooves	The hard nail-like growths covering the feet of buffalo, bison, and yak.
Mammals	The large group of animals that have hair on their body and feed their young on milk.
Mate	To come together to produce young.
Predators	Animals that hunt other animals.
Ruminating	Chewing the cud.
Species	The scientific word for animals of the same type that breed together.
Tuberculosis	A dangerous disease that affects the lungs.
Wallow	To rest or lounge around in.

Find Out More

Books

Jacobs, L. *Water Buffalo.* Wild Wild World. San Diego, California: Blackbirch Press, 2003.

Robbins, K. *Thunder on the Plains: The Story of the American Buffalo.* New York: Atheneum, 2001.

Web sites

Water Buffalo
animals.nationalgeographic.com/animals/mammals/water-buffalo.html
A profile of the water buffalo.

Yaks
www.enchantedlearning.com/subjects/mammals/cattle/Yakcoloring.shtml
Information about the yak and a printout to color in.

Index